YOUR KNOWLEDGE HAS VALUE

- We will publish your bachelor's and master's thesis, essays and papers

- Your own eBook and book - sold worldwide in all relevant shops

- Earn money with each sale

Upload your text at www.GRIN.com and publish for free

Frequency-domain analysis of discrete-time signals and design of Infinite-Impulse Response (IIR) and Finite-Impulse Response (FIR) filters

Bandar Hezam

Bibliographic information published by the German National Library:

The German National Library lists this publication in the National Bibliography; detailed bibliographic data are available on the Internet at http://dnb.dnb.de.

ISBN: 9783346817389
This book is also available as an ebook.

ASIA PACIFIC UNIVERSITY

TECHNOLOGY & INNOVATION

INDIVIDUAL LAB REPORT

FREQUENCY DOMAIN ANALYSIS & FILTER DESIGN

NAME: Bandar Naji Ali Hezam

SUBMISSION DATE: 17/9/2019

ACKNOWLEDGMENT

I would like to express my very great appreciation to **ASSOC. PROF. DR. THANG KA FEI** for his valuable and productive suggestions during the planning and development of this research work. **ASSOC. PROF. DR. THANG KA FEI** provided me with very valuable guidance and monitoring regarding the development and the design of the Filters. I would also like to extend my thanks to the technicians of the laboratory of APU for offering me the resources in running the program.

ABSTRACT

Digital filters provide an important role in the world of communication. This lab report will discuss the steps taken to develop and design IIR and FIR filters of two filters types which are the Low-pass and the Band-pass filters. The main objective of this lab report to compare and analyze the IIR and FIR filters. At first, a background about the topic of the report would be made. Then, the theoretical concept of the filters is being discussed. After that, the procedure and the steps taken to design the filters are to be explained. Then, the results would be displayed and analyzed. A discussion about the observations and findings made are to be stated. At last, a conclusion about the achievements made and the major outcomes of the investigation are stated.

OBJECTIVE

The main objectives of this lab report is:

1. Generate frequency-domain analysis of discrete-time signals.
2. Design appropriate Infinite-Impulse Response (IIR) and Finite-Impulse Response (FIR) filters.

TABLE OF CONTENTS

TABLE OF FIGURES

1 INTRODUCTION

One of the most dynamic areas of digital systems today is in the field of digital signal processing or DSP. A DSP is a very specialized form of microprocessor that has been optimized to perform repetitive calculations on stream of digitized data. (Dhar, 2019) The digitized data are usually being fed to the DSP from an ADC (Analog to Digital converter). A calculation is performed by the DSP to process these digitized data that come in. This calculation involves the most recent data point as well as several of the preceding data samples. The result of the calculation produces a new output data point, which is usually sent to a DAC (Digital to Analog) converter. A major application of DSP is in filtering and conditioning of analog signals. Filtering of a signal can be done by taking samples of the signal with an ADC, performing mathematical operation on the samples with a microcomputer and output the result to a DAC. This digital filter approach can easily produce filter response. The digital approach has the further advantage in that the filter response can be changed under program control.

Speech is the most basic and preferred means of communication among humans. In speech processing, a filter removes the unwanted signal and allows the desired signal. Filters may be analog or digital. Digital filtering is one of the important tools for digital signal processing applications. Digital filters can perform that specifications which are extremely difficult to achieve with an analog implementation. Multiple filtering is possible, and it can be operated over wide range of frequencies, because the characteristics of digital filters can be easily changed under software control. Digital filters are classified either as Finite duration impulse response (FIR) filters or Infinite duration impulse response (IIR) filters, depending on the form of impulse response of the system.

This assignment will include the development of a Lowpass filter and a Bandpass filter by using both IIR and FIR filters to a specific given signal. A comparison between both filters will then be made by comparing the achieved results based on the observations made.

2 THEORETICAL CONCEPTS

Detection of a wanted signal may be impossible if unwanted signals and noise are not removed sufficiently by filtering. Electronic filters allow some signals to pass but stop others. To be more precise, filters allow some signal frequencies applied at their input terminals to pass through to their output terminals with little or no reduction in signal level. An ideal filter can be described as a brick wall. However, this kind of filter does not exist in real world application. A further relationship between the time and frequency domains can be used to explain why the "brick wall" filter cannot exist. The reason why the "brick wall" filter cannot be built is because of the relationship between the time and frequency domains. Just as a voltage step function (a sudden change in the time domain) has frequency components that extend across a wide band, a step function in the frequency domain has voltage components that extend across a wide period of time. The frequency domain can be considered to cover both positive and negative frequencies, so a 1 kHz sine wave can be represented by a pair of spectral lines at +I kHz and -1 kHz. (Winder, 2002) The step frequency response will, by reciprocity, have time domain components at positive and negative time, relative to the event. Since a response cannot occur before an event has taken place (i.e., negative time), the step frequency response cannot exist.

An important relationship between the time domain and the frequency domain occurs when two signals are multiplied together. This relationship is important in digital filter design. Digital filters operate on digitized analog signals, so the digitization process is important and can be critical in the system design. Digitization requires the analog signal to be sampled and then converted into a digital value, based on the amplitude of the sample. A digital filter is a filter that works by performing digital mathematical operations on an intermediate form of a signal. It takes a digital input, gives a digital output and consists of digital components.

Digital filters are classified into two types depending on the duration of the impulse response. For finite-duration impulse response (FIR) digital filter, the operation is governed by linear constant-coefficient difference equations of a non-recursive nature. The transfer function of a FIR digital filter is a polynomial in z^{-1}. Infinite-duration impulse response (IIR) digital filters whose input-output characteristics are governed by linear constant-coefficient difference equations of a recursive nature. The transfer function of an IIR digital filter is a rational function in z^{-1}.

2.1 FINITE IMPULSE RESPONSE – FIR

A FIR filter comprises an array of delay elements connected in series. A tap is taken after each element, and, at any sample instance, the value of the sample is multiplied by a filter coefficient. Thus, a multiplier is needed for each delay element. Finally, the outputs of all the multipliers are added together to give the output. The number of taps is given by N, but there are N-1 delay elements; the term N-1 is sometimes referred to as the filter order. It is common to use an odd number of taps, which results in an even number of delay elements. Often, the filter coefficients are symmetrical. This allows us to design a hardware-reducing configuration where the delayed signal is fed back to halve the number of multipliers required. The circuit is folded around so that the first and last outputs from the delay line are added together and then multiplied by a common coefficient. Extra summing circuits are required, but the output stage adder has only half the number of inputs and therefore is simpler to implement. The duration or sequence length of the impulse response of these filters is finite. Therefore, the output can be written as a finite convolution sum by:

$$y[n] = \sum_{k=0}^{M} h[k]\, x[n-k]$$

FIR is designed by truncating the impulse-response of an ideal IIR filter. Truncating the envelope, by limiting its extent to a certain time limit, causes ripple in the frequency response passband and stopband, and limits the achievable stopband attenuation. Truncation can be applied gradually using specially designed window functions; these reduce the ripple effects and improve the stopband attenuation. Windows are applied by multiplying the window coefficients by the impulse response of the IIR filter. This would produce the impulse response of the FIR filter.

$$h[n] = h_d[n] \cdot w[n]$$

where h[n] is the impulse response of FIR filter, $h_d[n]$ is the impulse response of IIR filter and w[n] represents a rectangular window of length N. w[n] equals to one for $0 \leq n \leq N-1$ and zero otherwise. Different window functions can be used according to the application of the filter. Some of the most known window functions are Rectangular window, Bartlett window, Hamming window and Blackman window.

2.2 INFINITE IMPULSE RESPONSE – IIR

IIR filters are one of two primary types of digital filters used in Digital Signal Processing (DSP) applications (the other type being FIR). IIR filters are digital filters with infinite impulse response. Unlike FIR filters, they have the feedback (a recursive part of a filter) and are known as recursive digital filters. For this reason, IIR filters have much better frequency response than FIR filters of the same order. Digital filters with an Infinite-duration Impulse Response (IIR) have characteristics that make them useful in many applications. Infinite impulse response (IIR) filters are more efficient than FIR filters because, for a given frequency response, they require fewer delay elements, adders, and multipliers. The disadvantage of IIR filters is their nonlinear phase response. The design of an IIR filter can be done by following two steps. The first step is Analogue prototyping where the second step is Discretization. Most IIR filters are designed using an analog filter model. Analog filter models are the familiar Butterworth, Chebyshev, Cauer (Elliptic). Inverse Chebyshev, and Bessel types. The linear frequency response formulae H (ω) can be converted into the digital equivalent using Impulse Invariant, Step Invariant, or Bilinear Transformation. Only the bilinear transform provides a general-purpose conversion function that can be used for low-pass, high-pass, band-pass. and band-stop responses. The impulse invariant and step invariant conversion functions are quite difficult to apply and can only be used for low-pass filters (and band-pass with great care): these conversion functions cannot be used with high-pass or band-stop responses. The bilinear transform is used to convert the analog frequency response into a digital domain response. The advantage of the bilinear transform is that any response, be it low-pass, high-pass, band-pass, or band-stop, can be converted. The digital domain is also known as the Z-domain. The transformation from the analog S-plane into the digital Z-plane is quite simple to visualize. The S-plane frequency ($j\omega$) axis is wrapped around onto itself into the Z-plane to form a circle. One side of the circle is the zero-frequency point, which is the origin on the S-plane diagram. The other side of the circle is where the +infinity and -infinity points meet. The bilinear transform is a simple mathematical process. Starting with an analog frequency response, H(s), bilinear transformation to produce H(z) is made by substitution of s.

$$s = \frac{2}{T}\left(\frac{1 - z^{-1}}{1 + z^{-1}}\right), where\ T = \frac{1}{fs}$$

3 PROCEDURE

In this section, the steps taken to develop and design the IIR and FIR filters would be discussed. The filters were designed using SIMULINK in MATLAB software. Figure 1 shows the block diagram to be designed in order to develop the system. The system consists of four filters. The four filters are the IIR low-pass filter, the FIR low-pass filter, the IIR band-pass filter and the FIR band-pass filter. The input was gotten from the workspace and its settings were adjusted. Each of the block design filters were obtained from the DSP System Toolbox and their parameters were entered. The output was displayed in the spectrum analyzer. Each of the adjusted settings and all the parameters entered would be explained and showed in detail in this section.

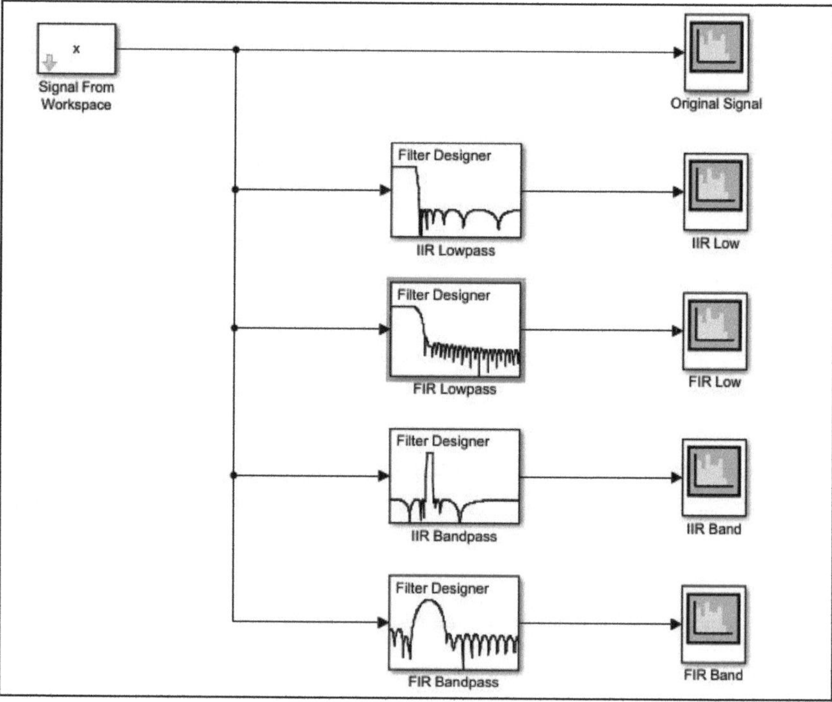

Figure 1 Main Block Diagram

At first, the signal given is loaded in the MATLAB software. From the loaded signal in the MATLAB, the sampling frequency is obtained. The obtained sampling frequency is 2000Hz. Also, the samples of the signal are obtained where there were 10,000 samples in the signal. Then, by using SIMULINK, the simulation stop time is made five seconds. This is made as there are 10,000 samples and the sampling frequency is 2000Hz. Therefore, a total of five seconds is needed. After that, a block, from the DSP System Toolbox, is entered. The block to be entered is the 'Signal from Workspace'. This block would allow the loaded signal from the MATLAB to be used in SIMULINK. To fully utilize this block, there are some parameters to be adjusted. The first parameter is to write the name of the signal as indicated in the workspace. The second parameter is the sampling time. The entered sampling time was 1/fs, as T, which is the sampling time is equal to 1/fs. At last, the number of samples per frame is entered. 512 samples per frame was chosen.

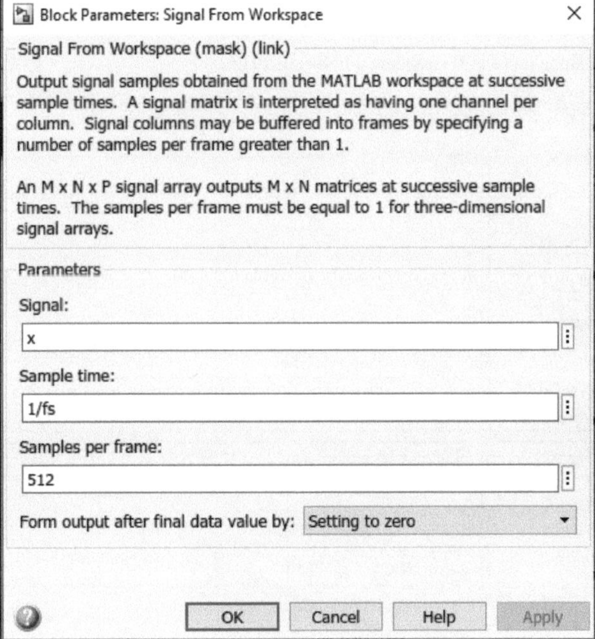

Figure 2 Signal from Workspace

After getting the signal into SIMULINK, a 'Digital Filter Design' block is to be inserted. This is to be inserted from the DSP System Toolbox. This block is to be used to design the filter. At first, two low-pass filter is to be developed where one of them is the IIR filter and the second is the FIR filter. The IIR low-pass filter is shown in Figure 3. There are some parameters to be changed to obtain the IIR low-pass filter. The first thing to be changed is the response type. Since a low-pass filter is to be designed, the response type is to be changed to lowpass. The design method to be used is the IIR and the analogue filter chosen was Chebyshev Type 2. This type of filter was chosen as it has a steep transition band comparing to other filters. Then, the filter method is to be specified as minimum order. This is specified as that to make the SIMULINK get the most accurate result using the minimum order possible. Then, the frequency specifications are to be specified. The fs is chosen as 2kHz. The passband frequency is chosen as 160 Hz where the stopband frequency is chosen as 220Hz. The passband frequency was chosen as that since the frequency component at 150Hz is chosen to be passed. The stopband frequency is chosen as such to make the order as less as possible and since there is no frequency before 220Hz.

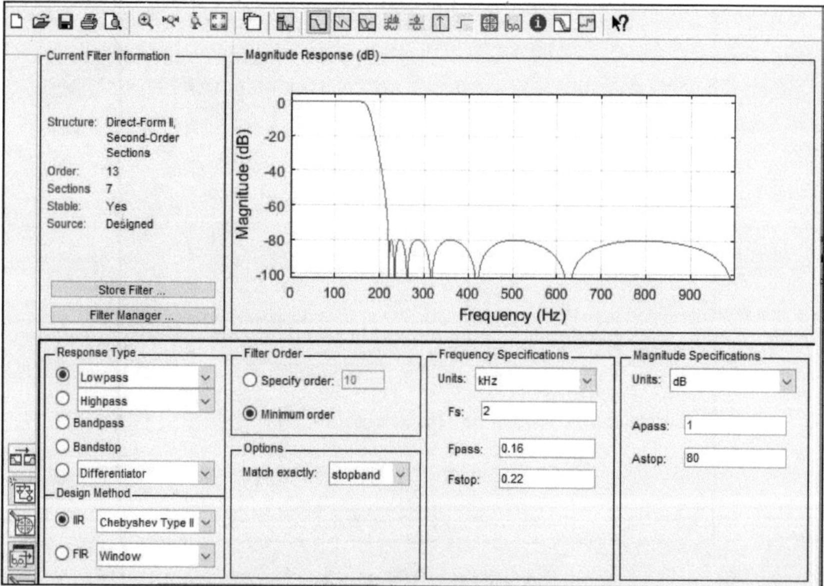

Figure 3 IIR Low-pass filter

After designing the low-pass IIR filter, the low-pass FIR filter is to be designed. The same block, which is 'Digital Filter Design' block, is to be inserted. This is to be inserted from the DSP System Toolbox. The parameters to be adjusted are shown in Figure 4. The response type to be chosen is the low-pass filter. Since a FIR filter is to be used, FIR is to be selected in the Design Method section. Also, in the dropdown menu, Window is to be chosen since FIR is designed using a window function. Then, an order is specified for the filter. The order specified was 60. This is chosen as 60 after several attempts. It is necessary to make the order as less as possible to make the complexity of the circuit less. However, to obtain an accurate result, the minimum order that can be used for this type of filter was 60. Then, a window function is chosen. Hamming window is to be chosen. This window was chosen where several window functions were used and the most accurate result with the least order obtained was by using the Hamming window function. Finally, the sampling frequency is entered as 2000Hz and the cut-off frequency is entered as 200Hz. It was entered as 200Hz so that the information is not lost, and the amplitude does not descend as the filter starts to enter the transition band. Therefore, a frequency which is slightly higher than the frequency which is desired to be kept is taken.

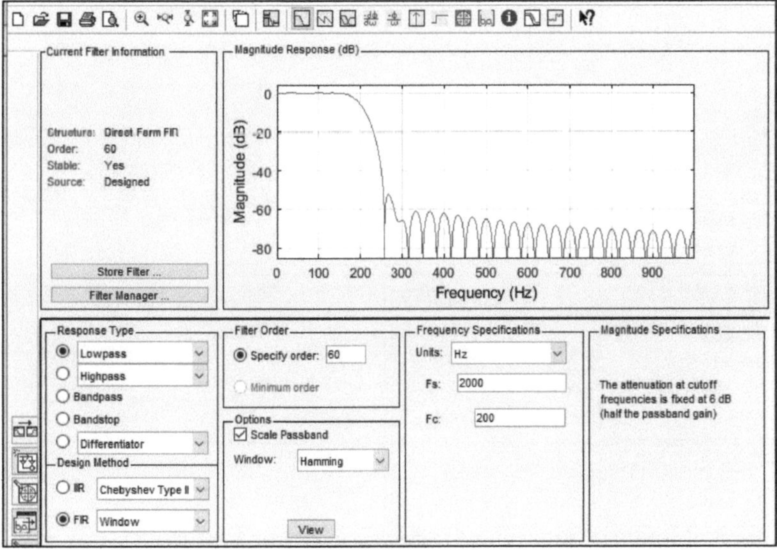

Figure 4 FIR Low-pass filter

The third filter to be designed is the Band-pass IIR filter. To do so, the response type is chosen as Band-pass. Then, the Chebyshev type 2 is chosen as the analogue filter for the IIR filter. This was chosen by first choosing IIR. Then, from the drop-down menu, the type of analogue filter is chosen. In this case, Chebyshev type 2 was chosen since it has a steep transition band comparing to other filters. After that, the filter order is chosen. In this option, the minimum order is to be chosen. By choosing this, MATLAB is given the freedom to get the most accurate result by having the least order. After that, the sampling frequency, fs, is inputted as 2000Hz. Then, the passband and stopband frequencies are entered. Since this is a bandpass filter. There are two stopband frequencies and two passband frequencies. The wanted frequency component is at 300Hz. Therefore, the first stopband frequency is made as 270Hz while the first passband frequency is made as 280Hz. The second passband frequency is made as 320Hz where the second stopband frequency is made is 330Hz. There was a difference made between the passband and the stopband frequencies to make the order less. If there was no difference between these frequencies, then the order would be high. If the order was high, then the filter would be complex to design.

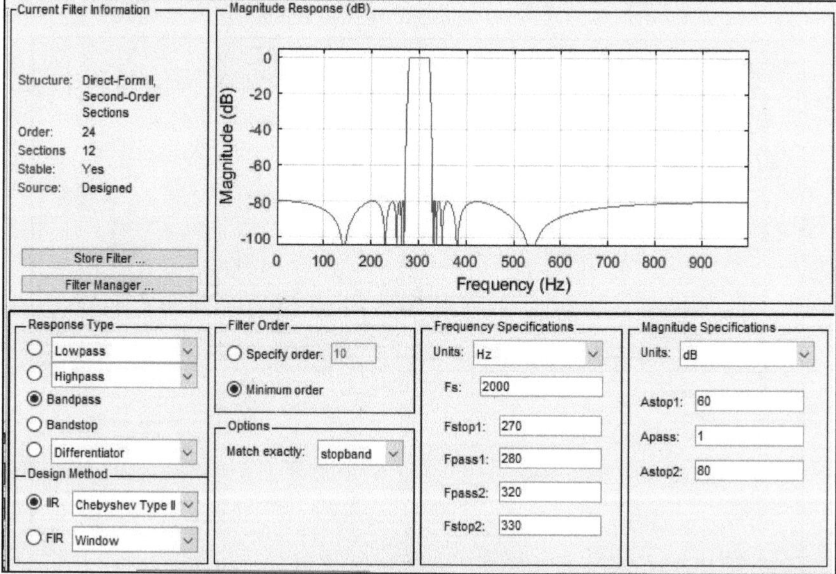

Figure 5 IIR Band-pass filter

The last and final filter to be designed is the FIR Band-pass filter. This was designed using the 'Digital Filter Design' block which was obtained from the DSP System Toolbox. The first parameter adjusted was in the Response type. This was made as Band-pass as this type of filter is needed to be developed. Since a FIR filter is to be designed, a window must be selected. To do so, in the 'Design Method' section, FIR is to be chosen and in the dropdown menu, the option 'Window' is to be chosen. Then, the order of the Filter is selected. To specify this, several attempts were made. It was found out that the most accurate result is obtained when the order is 30. However, better results can be obtained but the order of the filter would increase. This would make the filter design more complex to be designed. Then, the type of window function is selected. Hamming window was chosen. This window was chosen where several window functions were used and the most accurate result with the least order obtained was by using the Hamming window function. At last, the parameters for Fs, Fc1 and Fc2 are entered. These parameters are 2000Hz, 280Hz and 320Hz respectively. 2000Hz was chosen as it is the sampling frequency. 280Hz and 320Hz were chosen since the wanted frequency component is at 300Hz.

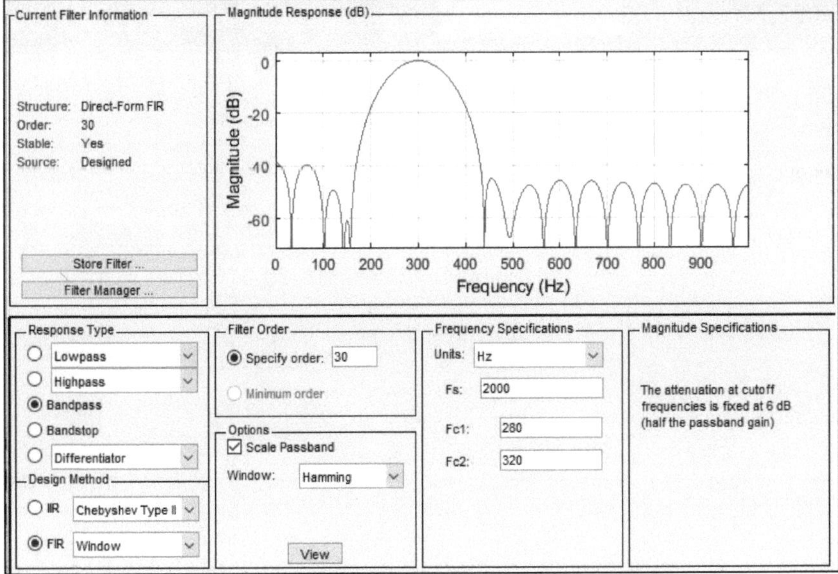

Figure 6 FIR Band-pass filter

At last and before testing the results of the developed filters, there were some settings which were needed to be adjusted in the Spectrum Analyzer. These settings are shown in Figure 7. The spectrum analyzer was obtained from the DSP System Toolbox. The first parameter to be adjusted is the input domain. Since the input is from Time domain, the input domain was changed in the dropdown menu to 'Time'. Then, the power spectrum of the signal is to be shown. Therefore, the type was changed to 'Power'. After that, in the 'Trace options', the units were changed to 'Watts'. Finally, only one side of the spectrum is to be displayed since the other side is a replica of the first side. So, analysis can be done on one-sided spectrum only. Therefore, the checkbox was unticked.

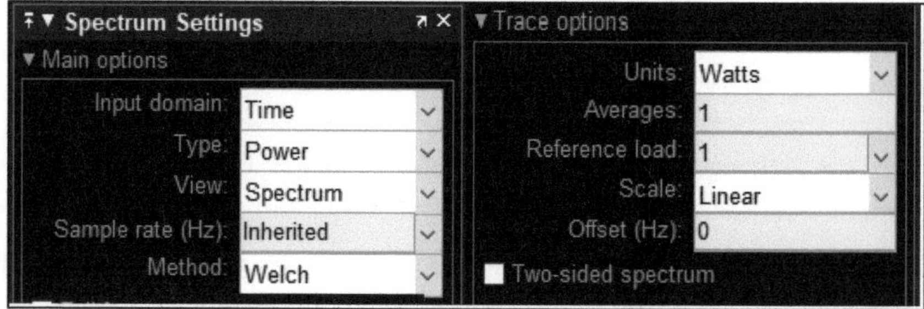

Figure 7 Spectrum Analyzer

4 ANALYSIS

This section will analyze the results gained and discuss each of the obtained plots. After developing the SIMULINK model where four filters were developed, the first result is the frequency domain of the original signal itself. This is shown in Figure 8. The graph shown in Figure 8 is the Power spectrum of the original signal. The x-axis is the frequency in kHz where the y-axis is the power in wats. As it can be seen from the figure, the original signal consists of five frequency components. The first frequency component lies at 150Hz and has a magnitude of 837.125mW. The Second frequency component lies at 220Hz and has a magnitude of 44.945mW. The third frequency component lies at 300Hz and has a magnitude of 1.994W. The fourth frequency component lies at 440Hz and has a magnitude of 403.092mW. The fifth frequency component lies at 660Hz and has a magnitude of 64.865mW

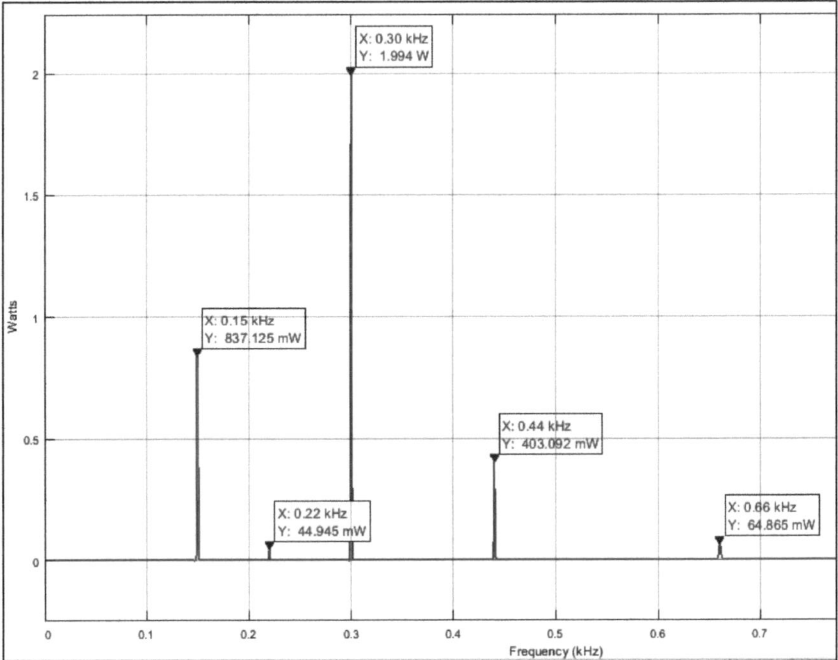

Figure 8 Original Signal

After analyzing the frequency domain of the original signal, the results that were gained after applying the filters are to be discussed. The first filter developed is the lowpass IIR filter. The frequency component that is to be kept is the one that has 150Hz. After designing the filter to that sort of specifications, the result was displayed in the spectrum analyzer. The result gained is shown in Figure 9. As it can be seen from the figure, all other frequency components were eliminated. The only frequency component remained is at 150Hz. However, this component had a minor loss in terms of its magnitude. The gained magnitude after implementing the filter is 836.985mW.

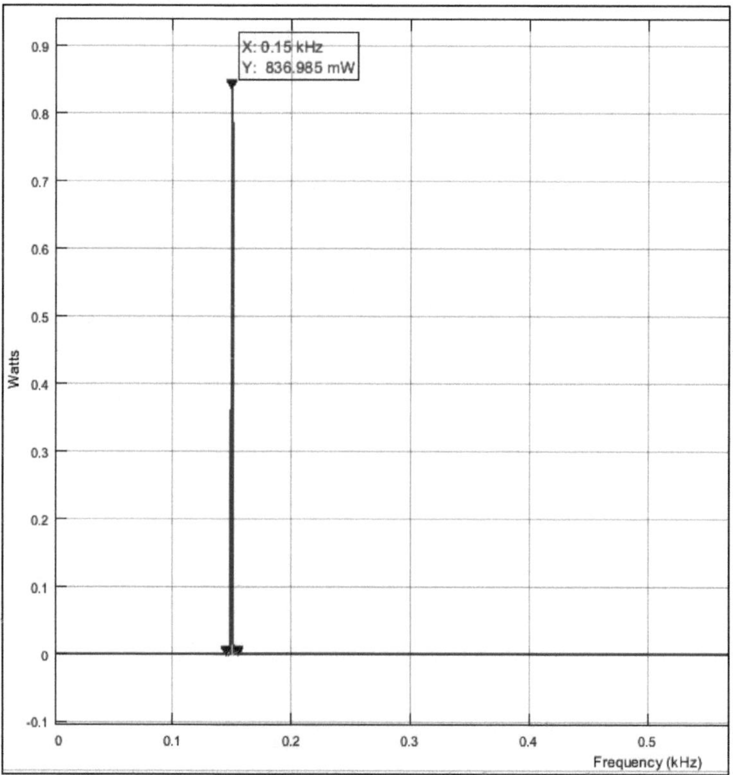

Figure 9 IIR Lowpass

The second filter that was designed is the FIR Low-pass filter. The same frequency component was kept for comparison purposes. This means that the frequency component at 150Hz is to be kept where others are to be eliminated. The result of this filter was displayed in a spectrum analyzer. The result gained is shown in Figure 10. It can be observed from figure 10 that all frequency components were eliminated except the one at 150Hz. However, there was an attenuation in terms of the magnitude of the component where the gained magnitude after implementing the filter is 825.187mW.

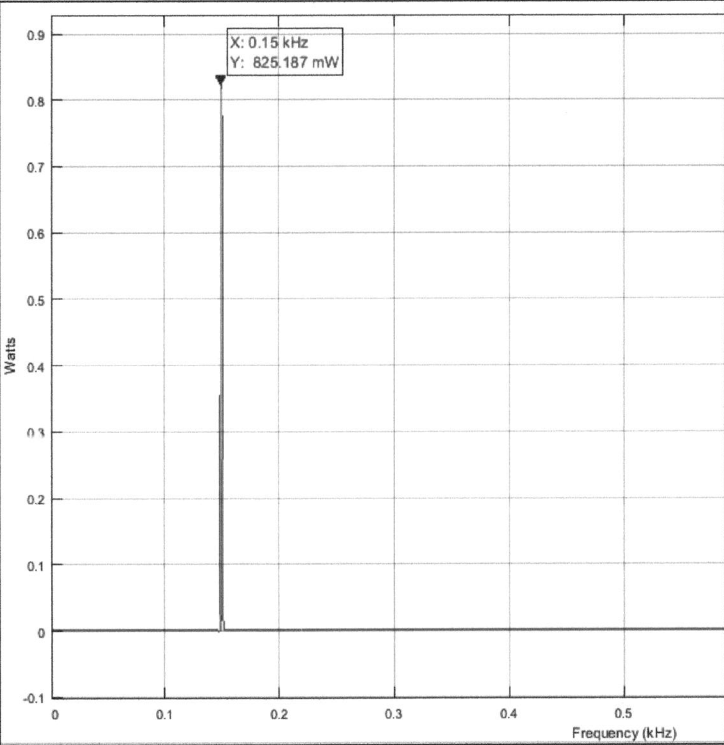

Figure 10 FIR Lowpass

The third filter that was designed is the IIR Bandpass filter. The purpose for this filter was to keep the frequency component in the middle, the one which has 300Hz. After designing the filter with these specifications, the result was displayed in the frequency spectrum. This is shown in Figure 11. It can be observed that the only frequency component remaining is the one with 300Hz. The magnitude for this frequency component is 1.994W.

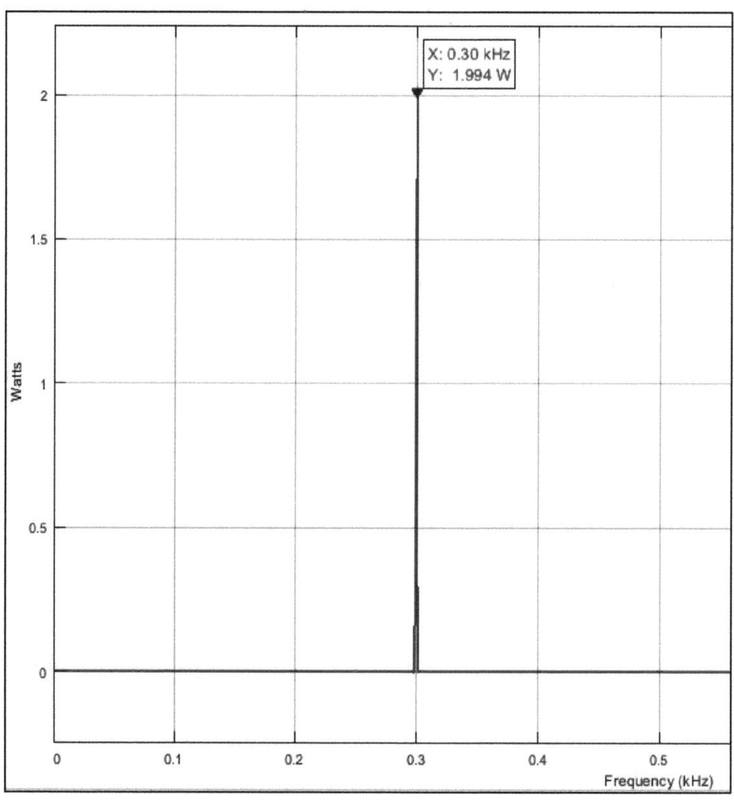

Figure 11 IIR Bandpass

The last filter designed is the FIR Bandpass filter. In this filter, the same frequency component that was kept in the IIR Bandpass filter is also kept. This was done to compare between the IIR and FIR filters. This means that the frequency competent with 300Hz is not be eliminated where other frequency components are eliminated. After getting the specifications for this filter and designing it, the result obtained is displayed in the spectrum analyzer. It can be seen from figure 12 that the only frequency component remained is the one with 300Hz. It has a magnitude of 1.994W.

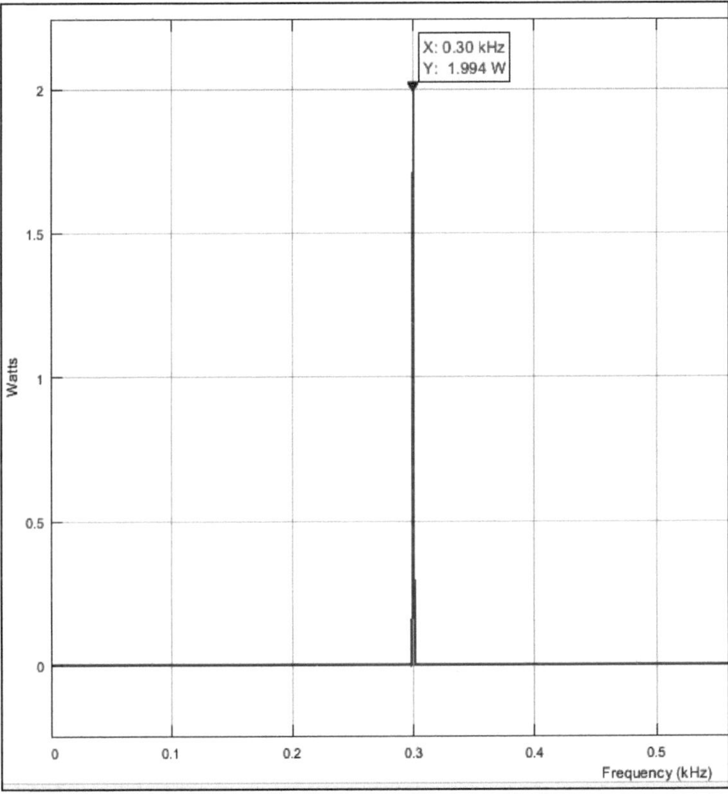

Figure 12 FIR Bandpass

After displaying the results for each filter, analysis on the results is made. Table 1 shows the filter order and the stability of all filters designed. From table 1, it can be seen that all filters designed were stable. This was said as the impulse response of all the filters are gradually decreasing to zero as the time increases. The stability of the IIR filters can also be determined by looking at the pole-zero plot of the filter. It was observed that all poles of all IIR filters are within the unit circle. However, in terms of the order of the filter, there were two lowpass filters developed and two bandpass filters. The two lowpass filters are of an IIR and a FIR filter. It was observed that the order of the IIR lowpass filter was 13 where the order of the FIR Lowpass filter is 60. The two bandpass filters are of an IIR and a FIR filter. The filter order for the bandpass is 24 where the filter order for the lowpass is 30. These filter orders were obtained to get the most accurate results.

Table 1 IIR and FIR

Filter type	Filter Order	Stable
IIR Lowpass	13	Yes
FIR Lowpass	60	Yes
IIR Bandpass	24	Yes
FIR Bandpass	30	Yes

One of the major differences between IIR and FIR filters is that FIR filters are always stable where stability cannot be always guaranteed in IIR filters. Also, FIR filters mostly require a higher filter order than IIR filters. Thus, require high computational efficiency, more memory and more processing time. Due to the high filter order, the FIR filters can also be more complex. In addition to, IIR can easily be developed where it is derived from Analogue filters. But this is not possible for FIR filters since they do not have analogue counterpart. However, since IIR have analogue counterpart, noise can easily be interfered. In this case, it can be observed that both filters, FIR and IIR, are stable. However, both FIR filters have a higher order than both IIR filters. This means that these filters require more computational efficiency, memory and processing time. From the result obtained, it can be observed that there was no noise interference in both IIR filters.

5 DISCUSSION

The process of developing and designing the filters started by making some researches about both IIR and FIR filters. It was required to get enough knowledge to start designing the filters based on their characteristics. After getting enough information, the filters were designed using SIMULINK in the MATLAB platform. The filters were designed, and the results were obtained. The original signal consisted of five different frequency components. A lowpass and bandpass filters were designed using IIR and FIR filters. The lowpass filters eliminated all frequency components except the 150 Hz. The bandpass filters eliminated all frequency components except the 300Hz. A comparison between IIR and FIR filters was made. However, there were some observations made.

One of the observations made was that the less the difference between the passband and the stopband frequencies in the IIR filter, which makes it act as a brick wall, the order of the filter increases. As the order of the filter increases, the accuracy of the result increases. However, this would also make it complex in terms of its design as it would have more delays. In addition to, it was observed that the magnitude of some frequency components was attenuated. This occurred because of the cut-off frequency which was set. For instance, the frequency component at 150Hz attenuated from 837.125mW to 825.187mW when passed by a FIR lowpass filter. The cut-off frequency was set at 200Hz. This means at 200Hz, the magnitude of the filter would be at -3dB. However, it starts decreasing before it reaches that value. This means that at 150Hz, the magnitude of the filter was not at 0dB, which is a gain of 1. It was slightly lower than that. Therefore, there was some attenuation observed in the frequency component.

Nevertheless, the result obtained were satisfactory and there were some differences between both IIR and FIR filters. But these differences lie in the terms of the magnitude of the signal not their frequencies. This was due to the cutoff frequencies that was set. However, the main difference that was noticed between the two filters were in terms of the filter order. The filter order of the FIR was relatively higher than the filter order of the IIR as expected. To enhance the results obtained, the filter order can be increased. This would make the result more accurate. However, it would also lead to some disadvantages. These disadvantages would be in terms of the implementation of the filter as it would be more complex. Since the signal given is small and the implementation is simple, IIR would be suitable as it is better for lower-order tapping.

6 CONCLUSION

To conclude, the objectives of the assignment have been successfully achieved where four filters were designed. Two of these filters were IIR filters and two of them are FIR filters. The filters were developed using the SIMULINK in MATLAB software. The results were obtained. A comparison between the IIR and FIR filters were made. It was found out that both filters are stable. However, the filter order of the FIR was more than the filter order of the IIR as expected. It was also observed that as the frequency is less between the passband and the stopband, the order of the filter is increased. As the order of the filter increases, the complexity of the filter also increases as more delays are added into the system. Moreover, it was also observed that some magnitudes of some frequency components are attenuated because of the cut-off frequency entered.

7 REFERENCES

Dhar, P. K., 2019. Design and Implementation of Non Real Time and Real Time Digital. Journal of Emerging Trends in Computing and Information Sciences, 2(3), pp. 149-155.

Hamming, R. W., 2013. Digital Filters. 2nd ed. New york: Dover Publications.

Schlichthärle, D., 2000. Digital Filters. 2nd ed. Germany: Springer-Verlag Berlin Heidelberg.

Winder, S., 2002. Analog and Digital Filter Design. 2nd ed. London: Newnes.